I0478040

Chapter 1: Introduction to Genetic Beauty

In an age where the allure of physical appearance plays a significant role in social dynamics, the quest for beauty has taken a profound turn. The exploration of **genetic beauty**—the idea that our DNA can dictate not only our health but also our aesthetic appeal—has emerged as a fascinating frontier. This chapter delves into how genetics influences physical appearance, the evolving societal perceptions of beauty, and the ethical considerations surrounding genetic modification.

Overview of Genetic Influence on Physical Appearance

From the moment we are conceived, our genetic blueprint begins to shape our physical form. The characteristics we inherit from our parents—ranging from skin color, hair texture, and eye shape to body structure—are all encoded within our genes. Recent advancements in genetics have revealed that these traits are not merely random; they are the result of complex interactions between multiple genes and environmental factors. Understanding this intricate relationship can empower individuals to embrace their unique genetic gifts while also exploring avenues for enhancement through gene editing.

The Societal Perception of Beauty and Its Evolution

Beauty standards have historically varied across cultures and eras, influenced by a myriad of factors including fashion, media representation, and societal norms. What was once deemed beautiful in one generation may not hold the same appeal in the next. In contemporary society, the rise of social media has amplified these changes, often creating unrealistic benchmarks for physical appearance. As such, there is an increasing desire for individuals to modify or enhance their attributes, sparking discussions about the role of genetics in achieving these ideals.

Moreover, the definition of beauty has expanded beyond superficial aesthetics to include health and vitality. Attributes such as clear skin, robust physique, and youthful appearance are often equated with overall well-being. This shift in perception prompts the question: can genetic editing be the key to unlocking a more universally attractive appearance?

Ethical Considerations in Genetic Modification

As we venture further into the realm of genetic enhancement, it is imperative to address the ethical implications associated with altering human genetics for aesthetic purposes. The prospect of modifying traits to achieve beauty raises questions about accessibility, equity, and the potential for societal pressure to conform to narrow beauty ideals.

While gene editing technologies like CRISPR have made significant strides in medicine, their application in cosmetic enhancements necessitates careful consideration. Who gets to decide what is attractive? Will these modifications create a new class of individuals with "desirable" traits, further exacerbating social inequalities?

In this chapter, we will not only explore the scientific basis for genetic beauty but also engage in thoughtful discussions about the moral responsibilities that come with such power. The intersection of beauty and genetics is not just a scientific endeavor; it is a reflection of our values, beliefs, and the very fabric of society.

As we continue through this book, we will unravel the intricacies of gene editing technologies, the science behind physical traits, and the implications for personal identity and societal norms. Our journey into **Genetic Charisma** will reveal not just the potential for enhanced beauty, but also the power and responsibility that comes with harnessing our genetic heritage. Together, let us explore how we can embrace advancements in genetics to cultivate not only an attractive appearance but also a deeper understanding of ourselves and our place in the world.

Chapter 2: The Science of Genetics

To fully appreciate the potential of **genetic beauty**, we must first delve into the fundamental concepts of genetics. This chapter will provide an overview of basic genetic principles, how genes influence our physical traits, and the intricate relationship between genetics, health, and beauty.

Basic Genetic Concepts and Terminology

At the core of genetics is the study of genes, the units of heredity that are passed from parents to offspring. Genes are segments of DNA (deoxyribonucleic acid) that contain the instructions for producing proteins, which are crucial for the structure and function of our bodies. Each gene can come in different versions known as alleles, which can affect the expression of traits.

For example, eye color is determined by several genes, each contributing to the overall phenotype—the observable characteristics of an individual. These traits can range from physical attributes like height, hair color, and skin tone to more complex characteristics like susceptibility to certain diseases.

Genotype refers to the genetic makeup of an individual, while **phenotype** is the physical expression of that genotype. Environmental factors, such as nutrition and exposure to toxins, can influence the phenotype, highlighting the complex interplay between genetics and environment.

How Genes Influence Physical Traits

The influence of genes on physical appearance is profound. From the moment of conception, an individual's genetic makeup begins to shape their characteristics. Specific genes have been identified that are directly linked to traits considered attractive, such as skin clarity, hair texture, and body proportions.

Research in the field of **epigenetics** reveals that genes can be turned on or off by various factors, including lifestyle choices and environmental exposures. This means that even if someone inherits genes associated with certain traits, the expression of those traits can be influenced by how they live their lives.

Furthermore, the concept of **genetic polymorphism**—the presence of multiple variations of a particular gene in a population—illustrates that diversity in genetics contributes to the wide range of appearances among individuals. Traits associated with beauty, such as facial symmetry and clear skin, can be traced back to genetic variations that are subject to both natural selection and cultural preferences.

The Role of Genetics in Health and Beauty

The connection between genetics and health cannot be overstated. Many physical traits that are deemed attractive are often indicators of underlying health. For instance, clear skin can reflect good health and vitality, while hair quality can be a sign of hormonal balance and nutritional status. This link means that beauty and health are intertwined; improving one can often lead to enhancements in the other.

Recent advancements in genetic research have uncovered specific genes linked to various health conditions, including those that can affect appearance. For instance, genes related to collagen production impact skin elasticity, which is crucial for maintaining a youthful appearance. Understanding these genetic factors provides insight into how we can potentially enhance our appearance through targeted interventions.

Moreover, with the rise of personalized medicine, individuals can gain access to genetic testing that informs them about their unique genetic predispositions. This knowledge can empower people to make informed decisions regarding their health and beauty routines, leading to optimized results.

In conclusion, genetics plays a pivotal role in determining physical traits that contribute to our perception of beauty. By understanding the fundamental concepts of genetics, we can begin to explore the potential for gene editing technologies to enhance these traits further. As we move forward in this book, we will delve into the groundbreaking advancements in gene editing and how they hold the key to unlocking a new standard of beauty that is not only attractive but also inherently healthy.

Chapter 3: Understanding Gene Editing Technologies

As we venture deeper into the realm of **Genetic Charisma**, it is essential to explore the tools and technologies that make genetic enhancement possible. This chapter will provide a comprehensive overview of gene editing technologies, with a particular focus on **CRISPR-Cas9**, the leading method in the field. We will also delve into the science behind gene modification and examine the current advancements that pave the way for transforming beauty through genetics.

Overview of CRISPR and Other Gene-Editing Techniques

CRISPR-Cas9 has revolutionized the field of genetics since its introduction in the early 2010s. This powerful technology allows for precise editing of DNA sequences within living organisms, enabling scientists to modify genes with remarkable accuracy and efficiency. The term "CRISPR" stands for "Clustered Regularly Interspaced Short Palindromic Repeats," a naturally occurring defense mechanism in bacteria that helps them fight viral infections. The system utilizes a guide RNA to locate a specific DNA sequence, and the Cas9 enzyme acts like a pair of molecular scissors, cutting the DNA at the targeted location.

Besides CRISPR, there are other gene-editing techniques worth mentioning:

1. **TALENs (Transcription Activator-Like Effector Nucleases)**: This method involves designing proteins that bind to specific DNA sequences and introduce double-strand breaks, which can lead to gene modifications as the cell repairs itself.
2. **ZFN (Zinc Finger Nucleases)**: Similar to TALENs, ZFNs use engineered proteins that can recognize and bind to specific DNA sequences. The binding triggers a break in the DNA, allowing for editing at that site.
3. **Meganucleases**: These are naturally occurring enzymes that can be engineered to target specific sequences of DNA, providing another method for gene editing.

Each of these technologies has its advantages and limitations, but CRISPR remains the most widely adopted due to its ease of use, cost-effectiveness, and versatility.

The Science Behind Gene Modification

The process of gene modification involves several steps:

1. **Identifying the Target Gene**: The first step is to identify which gene or genes are linked to the traits one wishes to enhance or modify. This could involve extensive research and data analysis.

2. **Designing the Guide RNA**: Once the target gene is identified, a guide RNA is designed to match the sequence of the gene. This RNA is crucial for directing the Cas9 enzyme to the correct location in the DNA.

3. **Introducing the CRISPR Components**: The guide RNA and Cas9 enzyme are introduced into the cells where gene editing is desired. This can be done using various delivery methods, such as viral vectors or electroporation.

4. **Cutting the DNA**: The Cas9 enzyme makes a cut at the targeted site in the DNA, resulting in a double-strand break.

5. **Repairing the DNA**: The cell's natural repair mechanisms kick in to fix the break. This repair process can lead to various outcomes:

- **Non-Homologous End Joining (NHEJ)**, which often results in insertions or deletions that can disrupt the gene function.

- **Homology-Directed Repair (HDR)**, where a donor DNA template can be provided to facilitate precise changes or corrections to the gene.

This powerful mechanism not only allows for modifications that can enhance beauty traits but also opens avenues for correcting genetic disorders that affect appearance.

Current Advancements in the Field of Genetic Editing

The field of gene editing is evolving rapidly, with ongoing research leading to innovative applications. Some of the notable advancements include:

- **Enhanced Precision**: Newer versions of CRISPR, such as CRISPR/Cas9 variants like CRISPR/Cas12 and CRISPR/Cas13, provide improved specificity and reduced off-target effects, making them safer for potential cosmetic applications.

- **Base Editing**: This innovative technique allows for the direct conversion of one DNA base pair into another, enabling precise edits without causing double-strand breaks. This can potentially reduce unwanted mutations and improve the reliability of genetic modifications.

- **Epigenome Editing**: This emerging area focuses on modifying the epigenetic marks that regulate gene expression without altering the underlying DNA sequence. It offers a way to influence traits without permanent changes to the genome, presenting a less invasive approach to enhancing beauty.

- **Clinical Trials**: Numerous clinical trials are underway exploring the safety and efficacy of gene editing techniques for various applications, including aesthetic enhancements. These trials aim to establish protocols that can eventually be used in cosmetic procedures.

In summary, understanding gene editing technologies is fundamental to unlocking the potential for enhanced attractiveness and allure through genetic modification. As we continue to explore the intersection of genetics and beauty, it is essential to remain informed about the science that drives these advancements and the ethical considerations that accompany them. The journey into the world of **Genetic Charisma** is not just about enhancing physical traits; it's about embracing the future of beauty through responsible and informed genetic enhancement.

Chapter 4: The Connection Between Genetics and Attractiveness

As we delve deeper into the intricate relationship between genetics and attractiveness, it's essential to define what attractiveness means from both biological and cultural perspectives. Our understanding of beauty is not solely determined by physical appearance but is also influenced by a complex interplay of genetic traits, societal standards, and personal experiences. This chapter will explore the genetic traits associated with perceived beauty, emphasizing the importance of factors such as symmetry, skin quality, and body shape.

Defining Attractiveness: Biological and Cultural Perspectives

Attractiveness can be broadly defined as the quality of being appealing or appealingly pleasant to look at, often associated with certain physical traits. Biologically, attractiveness has roots in evolutionary psychology, where certain characteristics signal health, fertility, and genetic fitness. Traits such as clear skin, bright eyes, and a symmetrical face have been universally recognized as indicators of good health and reproductive potential. These features are often seen as desirable across cultures, as they can suggest a person's ability to thrive and produce healthy offspring.

Culturally, the perception of beauty is shaped by societal norms, media representation, and historical context. What is deemed attractive in one culture may differ vastly in another. For instance, while certain body types may be celebrated in one region, they may be criticized in another. This dynamic nature of beauty standards highlights the role of culture in shaping individual perceptions of attractiveness.

Genetic Traits Associated with Perceived Beauty

Research in the field of genetics has revealed specific traits that are often correlated with perceived beauty. Some of these traits include:

1. **Facial Symmetry**: Numerous studies indicate that facial symmetry is a crucial component of attractiveness. Symmetrical faces are often perceived as more appealing because they are thought to reflect genetic stability and health. This preference for symmetry can be seen across various cultures, suggesting that it is a universal standard.

2. **Clear Skin**: Skin health plays a significant role in attractiveness. Clear, smooth skin is often associated with youthfulness and vitality, which are traits linked to reproductive fitness. Genetics can influence skin quality through factors such as melanin production, collagen levels, and susceptibility to skin conditions.

3. **Body Shape and Proportions**: Various body shapes have been idealized across cultures, but certain proportions—such as the waist-to-hip ratio in women and shoulder-to-waist ratio in men—are frequently associated with attractiveness. These ratios are influenced by genetics and hormonal factors, and they can signal fertility and strength, respectively.

4. **Hair Quality**: Hair type, color, and fullness can also contribute to perceived beauty. Genes determine hair texture and color, and traits such as shiny, healthy hair are often linked to overall attractiveness.

5. **Eye Color and Shape**: Eye color can evoke different perceptions of beauty and desirability. While preferences for eye color may vary culturally, certain traits—such as large, expressive eyes—are generally regarded as attractive. Genetic factors play a significant role in determining eye color and shape.

The Role of Symmetry, Skin Quality, and Body Shape in Attractiveness

The factors mentioned above—symmetry, skin quality, and body shape—interconnect to create an overall impression of beauty. Symmetry, for instance, is not just a standalone trait but interacts with other features to enhance attractiveness. A symmetrical face paired with clear skin can amplify an individual's appeal.

Skin quality is equally important, as it can either enhance or detract from other attractive traits. Flawless skin can accentuate symmetrical features and draw attention to a person's overall beauty. Conversely, skin conditions or blemishes may overshadow even the most symmetrical facial structure.

Body shape and proportions add another layer to the equation. The cultural ideals surrounding body shape are deeply ingrained, and societal pressures often dictate what is deemed attractive. Understanding the genetic underpinnings of these traits can help individuals embrace their natural beauty while considering enhancements through responsible gene editing.

Conclusion

The connection between genetics and attractiveness is multifaceted, encompassing both biological imperatives and cultural influences. By recognizing the traits commonly associated with beauty and understanding the genetic factors that contribute to these traits, we can begin to appreciate the profound interplay between our genes and our perception of self-worth. As we progress further into the realm of genetic modification, this understanding will provide a critical foundation for harnessing the potential of gene editing in pursuit of an attractive and alluring presence. The next chapter will delve into the ethical implications surrounding genetic enhancements and the debates that shape the conversation about beauty in the age of genetics.

Chapter 5: Ethical Implications of Genetic Enhancement

As we stand on the brink of revolutionary advancements in genetic editing and enhancement, the ethical implications surrounding these technologies demand our immediate attention. While the allure of using gene editing for cosmetic purposes is enticing, it raises fundamental questions about morality, equity, and the potential societal impacts of modifying human traits. In this chapter, we will explore the multifaceted debates surrounding genetic enhancement, examining both the potential benefits and the ethical dilemmas that accompany these advancements.

Debates Surrounding Gene Editing for Cosmetic Purposes

The possibility of using gene editing techniques like CRISPR to enhance physical appearance is both exciting and controversial. Proponents argue that genetic modifications could lead to improved health outcomes and the ability to attain beauty standards that promote self-esteem and social acceptance. By targeting genes responsible for traits such as skin quality, hair type, and body shape, individuals could potentially achieve their desired appearance.

However, critics raise concerns about the implications of treating beauty as a commodity. The pursuit of genetic enhancement for cosmetic purposes may perpetuate existing beauty standards, leading to increased pressure on individuals to conform to narrow definitions of attractiveness. This commodification of beauty could exacerbate societal inequalities, where only those who can afford genetic enhancements gain access to these perceived advantages, creating a new class of genetically "ideal" individuals.

Potential Societal Impacts of Genetic Modification

The societal impacts of genetic enhancement extend beyond individual choices and preferences. If cosmetic gene editing becomes widespread, it could shift cultural norms and values surrounding beauty. The pressure to modify one's appearance genetically may lead to a homogenization of beauty, where unique and diverse traits are overshadowed by a narrow set of ideals. This raises the question: what happens to the value of individuality in a world where genetic modifications are the norm?

Moreover, the implications of genetic enhancement could seep into other domains, including employment, relationships, and social interactions. Individuals who opt for genetic modifications may face biases or discrimination, while those who choose not to undergo such enhancements could find themselves marginalized. As beauty becomes increasingly intertwined with genetic status, the potential for social stratification based on genetic modifications becomes a pressing concern.

Responsible Use of Technology in Beauty Enhancement

To navigate the ethical landscape of genetic enhancement, it is crucial to establish guidelines and principles for the responsible use of these technologies. Here are some key considerations:

1. **Informed Consent**: Individuals considering genetic enhancements must be fully informed of the risks, benefits, and potential long-term consequences. Ensuring that patients understand the implications of gene editing is essential for ethical practice.

2. **Equity and Access**: As with any medical advancement, access to genetic enhancement technologies must be equitable. Efforts should be made to prevent a scenario where only the wealthy can afford enhancements while others are left behind. Policies should promote inclusivity and equal access to genetic advancements.

3. **Cultural Sensitivity**: Genetic enhancement practices must consider cultural diversity and the varying perceptions of beauty across different societies. Ethical considerations should take into account cultural values and avoid imposing a singular standard of attractiveness.

4. **Public Engagement**: Open discussions about the ethical implications of genetic enhancement should involve a broad range of stakeholders, including scientists, ethicists, policymakers, and the public. Engaging in dialogue about the potential consequences of genetic editing will help foster understanding and guide responsible use.

5. **Regulatory Frameworks**: Governments and regulatory bodies must establish clear guidelines and regulations governing the use of gene editing technologies for cosmetic purposes. These frameworks should prioritize safety, efficacy, and ethical considerations.

Conclusion

As we explore the potential of genetic editing to enhance physical appearance, it is imperative that we engage with the ethical implications that arise from these advancements. The pursuit of beauty through genetic modification must be approached with caution, balancing individual desires with societal considerations. By fostering responsible use and ethical guidelines, we can navigate the complex landscape of genetic enhancement while promoting inclusivity, individuality, and ethical standards in the pursuit of attractiveness.

In the following chapter, we will examine how the environment plays a crucial role in genetic expression, exploring the interplay between nature and nurture in the quest for beauty.

Chapter 6: The Role of Environment in Genetic Expression

As we delve deeper into the connection between genetics and attractiveness, it is crucial to recognize that our genes do not operate in a vacuum. While our genetic makeup provides the blueprint for our physical traits, the environment plays a vital role in shaping how those traits are expressed. This chapter will explore the interplay between nature and nurture, examining how various environmental factors can influence genetic potential and contribute to our overall appearance.

Nature vs. Nurture: How Environment Influences Genetic Potential

The longstanding debate of nature versus nurture encapsulates the essence of how our genetics interact with the environment. Nature refers to the biological inheritance we receive from our parents, comprising the genetic instructions that dictate our physical attributes. Nurture, on the other hand, encompasses the environmental influences and lifestyle choices that can affect how those genetic traits manifest.

Research has demonstrated that environmental factors such as diet, lifestyle, stress, and exposure to toxins can significantly influence gene expression. This field of study, known as epigenetics, explores how external factors can modify gene activity without altering the underlying DNA sequence. For example, certain diets can activate or deactivate specific genes related to skin health, hair growth, and body composition, demonstrating that our choices can optimize or inhibit our genetic potential.

The Impact of Lifestyle Choices on Physical Appearance

Our daily choices have a profound impact on our physical appearance, serving as an external influence on our genetic expression. Below are some key lifestyle factors that can enhance or detract from genetic traits associated with beauty:

1. **Diet**: Nutrition plays a critical role in maintaining healthy skin, hair, and body composition. Diets rich in antioxidants, vitamins, and healthy fats can promote radiant skin and robust hair growth. For instance, foods high in omega-3 fatty acids can improve skin hydration and elasticity, while antioxidants can protect against oxidative stress, a factor that contributes to premature aging.

2. **Physical Activity**: Regular exercise has been shown to improve body composition, muscle tone, and overall health. Engaging in physical activity not only supports cardiovascular health but also influences hormone levels, which can affect skin quality and body shape. Exercise can boost endorphins, improving mental well-being, which can further enhance one's appearance through increased confidence and reduced stress.

3. **Stress Management**: Chronic stress can have detrimental effects on physical appearance, leading to issues such as acne, hair loss, and dull skin. Implementing stress-reduction techniques, such as mindfulness, meditation, or yoga, can help mitigate these effects and promote a more attractive appearance.

4. **Sleep Quality**: Adequate sleep is essential for overall health and beauty. Sleep deprivation can lead to dark circles, dull skin, and increased signs of aging. Quality sleep allows the body to repair itself, supports hormonal balance, and enhances the skin's natural glow.

5. **Environmental Toxins**: Exposure to environmental pollutants, chemicals, and toxins can negatively affect skin health and overall appearance. Minimizing exposure to harmful substances and adopting a lifestyle that prioritizes clean living can help protect against damage to genetic expression and physical appearance.

Strategies for Optimizing Genetic Expression Through Environmental Factors

To harness the full potential of our genetics and enhance our physical presence, it is essential to adopt lifestyle practices that positively influence genetic expression. Here are some strategies to consider:

- **Nutritional Interventions**: Personalize your diet based on your genetic profile. Nutrigenomics can guide individuals in choosing foods that best support their unique genetic makeup, leading to optimized health and beauty.

- **Regular Exercise**: Create a balanced fitness regimen that includes cardiovascular, strength training, and flexibility exercises. Aim for at least 150 minutes of moderate-intensity aerobic activity per week, along with two or more days of strength training.

- **Stress Reduction Techniques**: Implement mindfulness practices, such as meditation or deep breathing exercises, to manage stress effectively. Incorporate activities that promote relaxation, such as yoga or nature walks, into your routine.

- **Prioritize Sleep**: Establish a regular sleep schedule, aiming for 7-9 hours of quality sleep each night. Create a calming bedtime routine to improve sleep quality, such as reducing screen time and maintaining a comfortable sleep environment.

- **Limit Exposure to Toxins**: Choose natural, non-toxic products for skincare and household cleaning. Advocate for cleaner air and water in your community to reduce environmental exposure.

Conclusion

Understanding the role of the environment in genetic expression is crucial for anyone seeking to enhance their physical appearance through genetic means. By recognizing the interplay between genetics and lifestyle choices, individuals can adopt practices that not only optimize their genetic potential but also promote overall well-being. In the following chapter, we will explore personalized beauty solutions through genetic testing, highlighting how genetic profiles can inform beauty treatments tailored to individual needs.

Chapter 7: Personalized Beauty: The Future of Genetic Testing

As we navigate the evolving landscape of beauty, one of the most exciting frontiers is the intersection of genetics and personalization. Genetic testing has emerged as a transformative tool, offering insights that can revolutionize how we approach beauty and health. This chapter will delve into the rise of genetic testing, how genetic profiles can inform tailored beauty treatments, and present case studies of personalized beauty solutions based on genetic information.

The Rise of Genetic Testing in Beauty and Health

In recent years, genetic testing has gained significant traction in both medical and cosmetic fields. With advancements in technology, it has become easier and more affordable for individuals to obtain their genetic information. This data can reveal a wealth of insights about predispositions to various health conditions, physical traits, and even aspects of beauty.

The application of genetic testing in beauty has taken on various forms, from at-home DNA kits to professional genetic consultations. These services analyze an individual's genetic markers to provide personalized recommendations tailored to their unique genetic profile. This shift towards customization reflects a broader trend in society where consumers increasingly seek products and solutions that cater to their specific needs and desires.

How Genetic Profiles Can Inform Beauty Treatments

Genetic profiles offer valuable information that can significantly enhance beauty and personal care routines. By understanding one's genetic makeup, individuals can make informed decisions about skincare, haircare, and overall wellness. Here are a few ways genetic profiles can inform beauty treatments:

1. **Skin Care Customization**: Genetic testing can reveal an individual's susceptibility to certain skin conditions, such as acne, eczema, or sensitivity to UV radiation. With this knowledge, personalized skincare regimens can be developed, targeting specific needs and reducing the risk of adverse reactions.

2. **Hair Care Solutions**: Genetic factors influence hair type, thickness, and the likelihood of hair loss. By analyzing genetic markers, individuals can receive tailored advice on haircare products and treatments that align with their hair profile, maximizing hair health and aesthetics.

3. **Nutritional Recommendations**: Nutritional genomics, or nutrigenomics, examines how genes affect individual responses to diet. By understanding their genetic predispositions, individuals can adopt dietary practices that support healthy skin and hair, improving their overall appearance from within.

4. **Tailored Exercise Plans**: Genetic profiles can also shed light on how individuals respond to different forms of exercise, which can influence physical appearance. Personalized fitness plans can be created based on genetic tendencies, optimizing results and enhancing physical attractiveness.

5. **Hormonal Health Insights**: Hormones play a crucial role in appearance, impacting everything from skin clarity to body composition. Genetic testing can provide insights into hormonal imbalances, guiding individuals toward treatments or lifestyle changes that promote hormonal health.

Case Studies of Personalized Beauty Solutions Based on Genetics

1. **23andMe and Skincare**: 23andMe, a well-known genetic testing company, has explored partnerships with skincare brands to create personalized skincare regimens based on genetic insights. Users can receive recommendations for products that suit their unique genetic predispositions, enhancing their skincare effectiveness and efficiency.

2. **Hair Color Customization with DNA**: Companies like Hairprint utilize genetic testing to offer personalized hair care solutions. By analyzing an individual's genetic data, these companies can recommend specific hair color products that work best with the individual's natural hair chemistry, resulting in healthier and more vibrant hair.

3. **NutriGenomics and Diet Plans**: Organizations specializing in nutrigenomics analyze genetic information to provide customized meal plans. By understanding how a person's body metabolizes certain nutrients, these plans can optimize skin health and promote an attractive appearance.

4. **Fitness DNA**: Fitness DNA is a service that provides insights into an individual's genetic predispositions related to physical performance and body composition. Users receive tailored exercise programs based on their genetic profiles, enhancing fitness outcomes and overall attractiveness.

Conclusion

The future of beauty lies in personalization, with genetic testing paving the way for tailored solutions that cater to individual needs. By harnessing the insights gained from genetic profiles, we can create beauty regimens that are not only effective but also promote confidence and self-acceptance. As we move forward, the integration of genetic information into beauty practices will likely become the norm, transforming our understanding of attractiveness and empowering individuals to embrace their unique genetic charisma.

In the following chapter, we will explore the fascinating world of nutrigenomics, delving into how our diets can enhance our genetic potential and promote an attractive appearance.

Chapter 8: Nutrigenomics: Eating for Your Genes

As we delve deeper into the relationship between genetics and beauty, we encounter a fascinating field known as nutrigenomics. This emerging discipline examines how our genes interact with the nutrients we consume, revealing the profound impact diet can have on our genetic expression and overall appearance. This chapter will explore the connection between diet and gene expression, identify foods that enhance beauty at a genetic level, and outline nutritional strategies for promoting an attractive appearance.

The Connection Between Diet and Gene Expression

Nutrigenomics posits that our dietary choices can significantly influence the way our genes function. Each individual's genetic makeup can affect how their body metabolizes nutrients, which in turn impacts various aspects of health, including physical appearance. For instance, certain genetic variations may predispose individuals to respond differently to specific nutrients, leading to variations in skin health, hair quality, and even body composition.

Research has shown that specific dietary components can activate or silence particular genes—processes known as gene expression regulation. For example, antioxidants found in fruits and vegetables can promote the expression of genes associated with skin health, while unhealthy fats can trigger inflammatory responses, negatively affecting both health and appearance.

Foods That Enhance Beauty at a Genetic Level

1. **Antioxidant-Rich Foods**: Foods high in antioxidants, such as berries, nuts, and dark leafy greens, play a crucial role in protecting our cells from oxidative stress. Antioxidants can help maintain skin elasticity and reduce signs of aging by supporting the expression of genes involved in skin repair and regeneration.

2. **Omega-3 Fatty Acids**: Found in fatty fish like salmon, walnuts, and flaxseeds, omega-3 fatty acids are essential for maintaining healthy skin and hair. They help regulate inflammation and can positively influence the expression of genes related to skin hydration and barrier function.

3. **Vitamin C Sources**: Citrus fruits, strawberries, and bell peppers are excellent sources of vitamin C, which is vital for collagen synthesis. Collagen is a protein that provides structure to our skin and helps maintain its firmness and elasticity. By ensuring adequate vitamin C intake, individuals can promote the expression of collagen-related genes.

4. **Polyphenol-Rich Foods**: Foods such as green tea, dark chocolate, and red wine are rich in polyphenols, which have been shown to have protective effects on skin health. These compounds can modulate gene expression related to inflammation and oxidative stress, contributing to a more youthful appearance.

5. **Lean Proteins**: Protein is essential for tissue repair and regeneration. Consuming lean proteins, such as chicken, turkey, legumes, and plant-based protein sources, can support muscle health and contribute to a toned appearance. Proteins also play a vital role in the production of enzymes and hormones that regulate various bodily functions.

Nutritional Strategies for Promoting an Attractive Appearance

To harness the power of nutrigenomics and enhance genetic expression for beauty, consider the following strategies:

1. **Balanced Diet**: Aim for a balanced diet that includes a variety of nutrient-dense foods. A colorful plate filled with fruits, vegetables, whole grains, lean proteins, and healthy fats will provide a broad spectrum of vitamins and minerals essential for optimal health and appearance.

2. **Hydration**: Staying adequately hydrated is crucial for maintaining healthy skin. Water aids in digestion, nutrient absorption, and toxin elimination. Aim to drink at least eight glasses of water a day, adjusting based on individual needs and activity levels.

3. **Mindful Eating**: Practicing mindfulness during meals can enhance the enjoyment of food and promote better digestion. Being present while eating allows individuals to appreciate the flavors, textures, and nutritional value of their meals, fostering a more positive relationship with food.

4. **Personalized Nutrition**: Consider consulting with a nutritionist or healthcare professional who can provide personalized dietary recommendations based on genetic testing. By understanding how your unique genetic profile influences your dietary needs, you can make informed choices that support your health and beauty goals.

5. **Supplement Wisely**: While obtaining nutrients through food is ideal, supplements can be beneficial in some cases, especially if deficiencies are present. Consult with a healthcare professional before starting any supplementation to ensure it aligns with your individual health needs.

Conclusion

Nutrigenomics represents a powerful intersection between our diets and our genetic potential. By understanding how food influences gene expression, we can make informed dietary choices that enhance our beauty and overall health. Incorporating nutrient-rich foods that promote genetic expression will not only contribute to an alluring appearance but also support long-term well-being. In the next chapter, we will explore the psychological aspects of beauty, including how self-perception and societal standards influence our understanding of attractiveness.

Chapter 9: The Psychology of Beauty

Beauty is not merely a visual experience; it is deeply intertwined with psychology, culture, and personal identity. Understanding the psychology of beauty is crucial in recognizing how self-perception influences attractiveness, the psychological benefits of enhancing physical appearance, and the impact of societal standards on individual beauty. This chapter delves into these dimensions, exploring how our perceptions and experiences shape our understanding of beauty.

How Self-Perception Influences Attractiveness

Self-perception refers to how individuals view themselves, which can significantly impact their confidence and the way they present themselves to the world. When individuals possess a positive self-image, it often reflects in their demeanor, body language, and interactions with others, enhancing their attractiveness. Conversely, a negative self-perception can lead to self-doubt and a lack of confidence, detracting from one's overall appeal.

1. **The Role of Confidence**: Confidence is a critical component of perceived attractiveness. Research shows that individuals who exude confidence are often viewed as more attractive, regardless of conventional beauty standards. This phenomenon is supported by the "halo effect," where positive traits, such as charisma and self-assuredness, enhance an individual's appeal.

2. **Self-Care and Self-Perception**: Engaging in self-care routines—whether through physical fitness, skincare, or personal grooming—can improve self-esteem and foster a more positive self-image. As individuals invest time and effort into their appearance, they often experience a boost in confidence, which can make them more attractive to others.

3. **Mental Health and Beauty**: Mental health plays a pivotal role in self-perception. Conditions such as anxiety and depression can distort one's view of themselves and diminish confidence, leading to a negative spiral regarding physical appearance. Conversely, individuals who seek mental health support may find improved self-acceptance and a healthier body image, thereby enhancing their attractiveness.

The Psychological Benefits of Enhancing Physical Appearance

Enhancing physical appearance through various means—such as beauty products, cosmetic procedures, or even genetic modification—can yield significant psychological benefits:

1. **Increased Self-Esteem**: Enhancements can lead to greater self-esteem and confidence. When individuals feel satisfied with their appearance, they are more likely to engage socially, take risks, and pursue opportunities they may have previously avoided.

2. **Empowerment**: The ability to make choices about one's appearance can foster a sense of empowerment. Individuals who actively participate in shaping their looks may feel more in control of their lives and identities.

3. **Social Acceptance**: While it is vital to challenge societal standards of beauty, there is no denying that conforming to certain aesthetic norms can lead to social acceptance and positive interactions. For some, this acceptance can facilitate personal and professional opportunities, contributing to overall well-being.

4. **Therapeutic Effects**: Engaging in beauty enhancement can have therapeutic effects, providing a sense of accomplishment and satisfaction. The process of self-improvement, whether through fitness, skincare, or cosmetic procedures, can serve as a form of self-care and emotional healing.

The Impact of Societal Standards on Individual Beauty

Societal standards of beauty can significantly influence individual perceptions of attractiveness. These standards, shaped by culture, media, and historical context, dictate what is considered beautiful or desirable, often leading to unrealistic expectations.

1. **Media Influence**: The portrayal of beauty in media—through advertisements, social media, and television—can create a narrow definition of attractiveness that many feel pressured to achieve. This can lead to dissatisfaction and body image issues, as individuals compare themselves to idealized images.

2. **Cultural Variations**: Beauty standards can vary significantly across cultures. What is deemed attractive in one culture may not hold the same value in another. Understanding these variations can promote a more inclusive and diverse appreciation of beauty.

3. **Challenging Norms**: The rise of movements advocating for body positivity and inclusivity has prompted a reevaluation of beauty standards. Individuals are increasingly encouraged to embrace their unique attributes, promoting a broader understanding of attractiveness that transcends conventional norms.

4. **Social Media's Double-Edged Sword**: While social media can perpetuate unrealistic beauty standards, it also serves as a platform for diversity and representation. Users can find communities that celebrate varied definitions of beauty, fostering acceptance and self-love.

Conclusion

The psychology of beauty is a complex interplay of self-perception, societal standards, and personal choices. Understanding how these elements shape our views can empower individuals to embrace their unique attractiveness and challenge the often-restrictive norms imposed by society. As we continue our exploration of genetic charisma, we will delve into the connection between genetics and the beauty industry, revealing how scientific advancements can align with our evolving perceptions of attractiveness.

Chapter 10: Genetic Disorders and Beauty

In the quest for beauty, it is essential to recognize that genetics plays a complex role in shaping our physical appearance. While many aspire to enhance their attractiveness through genetic editing and modification, we must also consider the implications of genetic disorders that can affect beauty. This chapter explores common genetic disorders that impact appearance, the beauty challenges associated with these conditions, and the importance of education and awareness in fostering a more inclusive understanding of beauty.

Common Genetic Disorders That Affect Appearance

Genetic disorders arise from abnormalities in an individual's DNA, which can lead to various physical characteristics and health issues. Some of these conditions can significantly alter a person's appearance, influencing how they are perceived by society. Here are a few common genetic disorders that affect beauty:

1. **Down Syndrome**: Individuals with Down syndrome often exhibit distinct facial features, including a flat facial profile, slanted eyes, and a shorter stature. While these traits are characteristic of the condition, it is vital to celebrate the unique beauty and individuality of each person, moving beyond stereotypes.

2. **Marfan Syndrome**: This disorder affects connective tissues, leading to a tall and slender physique, long limbs, and long fingers. Individuals with Marfan syndrome may also experience various health issues, such as cardiovascular problems. The emphasis on height and slenderness can create societal pressures, highlighting the need for a more nuanced understanding of beauty.

3. **Ehlers-Danlos Syndrome (EDS)**: EDS is a group of disorders affecting connective tissue, leading to hyper-flexibility, fragile skin, and a range of other symptoms. People with EDS may face challenges in their appearance due to skin fragility, but many embody resilience and strength that goes beyond conventional beauty standards.

4. **Albinism**: Albinism results in little or no pigmentation in the skin, hair, and eyes, leading to distinctive physical traits. While individuals with albinism often face societal stigma and challenges related to sun sensitivity and vision issues, embracing their unique appearance is essential for promoting inclusivity in beauty.

Understanding and Addressing Beauty Challenges Associated with Genetic Conditions

Individuals with genetic disorders may encounter specific beauty challenges that can affect their self-esteem and social interactions. It is crucial to approach these challenges with empathy and understanding:

1. **Social Stigma**: People with genetic disorders often face stigma and discrimination due to societal perceptions of beauty. These negative attitudes can lead to isolation, bullying, and mental health challenges. Promoting awareness and understanding is vital to combat these issues and foster inclusivity.

2. **Self-Image and Confidence**: Many individuals with genetic disorders struggle with self-image, especially in cultures that prioritize conventional beauty standards. Support systems, therapy, and community engagement can help individuals build confidence and self-acceptance.

3. **Accessible Beauty Products**: The beauty industry often overlooks the needs of individuals with genetic disorders. Creating inclusive beauty products that cater to diverse skin tones, hair types, and conditions can empower individuals to express their unique beauty.

4. **Representation in Media**: Increasing representation of individuals with genetic disorders in media and advertising can challenge stereotypes and promote a broader definition of beauty. By showcasing diverse appearances, society can shift its perception and embrace beauty in all forms.

The Importance of Education and Awareness

Education and awareness are crucial in shaping societal attitudes toward genetic disorders and beauty. By fostering understanding, we can create an environment that values diversity and promotes acceptance:

1. **Public Awareness Campaigns**: Initiatives aimed at educating the public about genetic disorders can help dispel myths and misconceptions, fostering a culture of empathy and acceptance.

2. **Advocacy and Support Groups**: Organizations dedicated to supporting individuals with genetic disorders can provide valuable resources, connect individuals with similar experiences, and advocate for policies that promote inclusivity.

3. **Collaborative Research**: Investing in research to better understand genetic disorders and their implications for beauty can lead to innovations in both genetic editing and beauty products, ultimately enhancing the quality of life for those affected.

Conclusion

While genetics plays a significant role in shaping beauty, it is vital to recognize the complexities associated with genetic disorders. By understanding the challenges faced by individuals with these conditions, we can foster a more inclusive definition of beauty that celebrates diversity. The journey toward genetic charisma involves not only enhancing physical traits but also embracing the unique qualities that make each individual beautiful. As we continue our exploration of genetic editing and beauty, let us remain committed to promoting understanding, acceptance, and inclusivity in all its forms.

Chapter 11: The Role of Hormones in Attractiveness

Hormones are powerful chemical messengers that play a significant role in regulating various bodily functions, including those that contribute to physical appearance and attractiveness. In this chapter, we will explore how hormonal balance affects physical traits, the impact of hormones on skin, hair, and body composition, and strategies for hormonal optimization that can enhance one's allure.

How Hormonal Balance Affects Physical Traits

Hormones influence the development of secondary sexual characteristics, which can significantly affect perceptions of attractiveness. For example, levels of estrogen and testosterone are critical in determining body fat distribution, muscle mass, and overall physical stature, all of which contribute to attractiveness as defined by societal standards.

1. **Estrogen**: This hormone is primarily responsible for the development of female secondary sexual characteristics. It promotes the growth of breast tissue, influences fat distribution (leading to a more rounded silhouette), and enhances skin hydration. High levels of estrogen are often associated with youthfulness and fertility, traits that are universally considered attractive.

2. **Testosterone**: Known as the male sex hormone, testosterone plays a key role in developing male secondary characteristics, such as increased muscle mass, facial hair, and deeper voice. In both men and women, adequate levels of testosterone are linked to a higher sex drive and overall vitality, which can enhance attractiveness.

3. **Thyroid Hormones**: The thyroid gland produces hormones that regulate metabolism. An imbalance in these hormones can lead to weight gain or loss, affecting body shape and size. A healthy thyroid function is essential for maintaining energy levels and a desirable physique.

4. **Cortisol**: Often referred to as the "stress hormone," cortisol can have negative effects on physical appearance when present in excess. Chronic stress leading to high cortisol levels can contribute to weight gain, particularly around the abdomen, and can also lead to skin issues such as acne or premature aging.

The Impact of Hormones on Skin, Hair, and Body Composition

Hormones play a vital role in determining skin quality, hair health, and body composition, all of which contribute to overall attractiveness.

1. **Skin Quality**: Hormones like estrogen and progesterone help maintain skin elasticity and hydration. As these hormone levels fluctuate, particularly during puberty, menstruation, or menopause, individuals may experience changes in skin quality, such as increased oiliness, acne, or dryness. Proper hormonal balance can lead to clearer, healthier skin, enhancing attractiveness.

2. **Hair Health**: Hormones also influence hair growth and quality. For instance, testosterone can lead to increased hair growth in areas like the face and body for men, while high estrogen levels can promote fuller hair for women. Conditions like polycystic ovary syndrome (PCOS), characterized by hormonal imbalances, can lead to hair thinning or excessive body hair, impacting self-esteem and perceptions of beauty.

3. **Body Composition**: Hormones are integral to regulating body fat and muscle mass. For instance, anabolic hormones like growth hormone and testosterone promote muscle development, while estrogen affects fat distribution, particularly during reproductive years. A balanced hormonal profile can lead to a more attractive body composition, which is often associated with health and vitality.

Strategies for Hormonal Optimization

Optimizing hormone levels can enhance attractiveness and overall health. Here are several strategies to consider:

1. **Diet and Nutrition**: Consuming a balanced diet rich in whole foods, healthy fats, and lean proteins can support hormonal balance. Certain nutrients, like omega-3 fatty acids, zinc, and vitamin D, play essential roles in hormone production and regulation.

2. **Regular Exercise**: Physical activity, particularly strength training and cardiovascular exercise, can help regulate hormones like testosterone and cortisol. Exercise has also been shown to improve mood and reduce stress, which positively influences hormone levels.

3. **Adequate Sleep**: Sleep is crucial for maintaining hormonal balance. Insufficient sleep can lead to increased cortisol levels and hormonal disruptions. Aiming for 7-9 hours of quality sleep per night can support overall health and beauty.

4. **Stress Management**: Reducing stress through techniques such as mindfulness, yoga, or meditation can lower cortisol levels and help balance other hormones. Finding healthy outlets for stress can improve both mental and physical attractiveness.

5. **Consultation with Healthcare Professionals**: For individuals experiencing significant hormonal imbalances, consulting with healthcare professionals, including endocrinologists or nutritionists, can provide personalized strategies and potential treatment options to optimize hormonal health.

Conclusion

Hormones play a crucial role in defining attractiveness through their influence on physical traits such as skin quality, hair health, and body composition. By understanding and optimizing hormone levels, individuals can enhance their physical appearance and overall well-being. As we continue to explore the connections between genetics, beauty, and health, acknowledging the role of hormones will be vital in harnessing genetic charisma and achieving an alluring presence.

Chapter 12: Beauty and Aging: A Genetic Perspective

As we traverse the journey of life, aging is an inevitable process that affects everyone. This chapter delves into the genetic underpinnings of aging and their direct implications on beauty and physical appearance. By exploring the genetic factors that influence how we age, current advancements in anti-aging gene therapies, and lifestyle changes that can help maintain a youthful appearance, we aim to provide a comprehensive understanding of beauty in the context of aging.

The Genetics of Aging and Its Effects on Appearance

Aging is influenced by both genetic and environmental factors. Geneticists have identified several key genes associated with the aging process, which can impact various aspects of physical appearance, including skin elasticity, hair loss, and the appearance of wrinkles.

1. **Telomeres**: These protective caps at the ends of chromosomes shorten as cells divide. Over time, telomere shortening is linked to cellular aging and the onset of age-related physical changes. Individuals with longer telomeres may experience a slower aging process, potentially maintaining a more youthful appearance longer.

2. **Sirtuins**: A family of proteins that play a critical role in cellular regulation, including aging, inflammation, and stress resistance. Studies suggest that sirtuins can promote DNA repair and improve cellular health, thereby influencing the aging process.

3. **The FOXO Gene**: This gene has been associated with longevity and the regulation of metabolism, oxidative stress, and apoptosis (cell death). Variants of the FOXO gene can affect the aging process, leading to differences in how individuals experience aging.

4. **Collagen Production Genes**: Collagen is a crucial protein that helps maintain skin elasticity and structure. Genetic variations can affect how much collagen is produced and the quality of the collagen itself, influencing the appearance of wrinkles and skin sagging as we age.

Current Research on Anti-Aging Gene Therapies

Recent advancements in genetic research have opened new avenues for anti-aging therapies. The focus is on using gene editing technologies to target the genes associated with aging and enhance the body's natural ability to repair and regenerate.

1. **Gene Editing for Skin Regeneration**: Emerging studies are exploring the potential of using gene editing tools like CRISPR to enhance collagen production in skin cells. By targeting specific genes responsible for collagen synthesis, researchers aim to develop treatments that can reduce wrinkles and improve skin texture.

2. **Gene Therapies for Hair Regrowth**: Age-related hair loss is a common concern for many individuals. Genetic therapies targeting the genes associated with hair follicle health are being investigated to promote hair regeneration and restore natural hair density.

3. **Telomere Extension Therapies**: While still largely experimental, researchers are exploring ways to lengthen telomeres in human cells. The idea is that extending telomeres could delay the onset of age-related cellular degeneration, potentially leading to a more youthful appearance.

Lifestyle Changes that Promote Youthful Looks

While genetic factors play a crucial role in aging, lifestyle choices can significantly impact how those genetic predispositions manifest. Here are several strategies that can help maintain a youthful appearance:

1. **Nutrition**: A balanced diet rich in antioxidants, vitamins, and healthy fats can combat oxidative stress, a key factor in aging. Foods such as berries, nuts, avocados, and leafy greens can support skin health and overall well-being.

2. **Physical Activity**: Regular exercise has been shown to improve circulation, enhance skin tone, and promote overall health. Engaging in both cardiovascular and strength-training exercises can help maintain muscle mass and combat age-related weight gain.

3. **Sun Protection**: Ultraviolet (UV) radiation is a significant contributor to premature aging of the skin. Using sunscreen daily and seeking shade when outdoors can help protect against skin damage and maintain a youthful appearance.

4. **Stress Management**: Chronic stress can accelerate the aging process. Techniques such as mindfulness, meditation, and yoga can help reduce stress levels and promote mental well-being, which can positively influence physical appearance.

5. **Quality Sleep**: Sleep is essential for cellular repair and regeneration. Prioritizing good sleep hygiene and ensuring adequate sleep can improve skin health, reduce dark circles, and enhance overall vitality.

Conclusion

The intersection of genetics and aging presents a fascinating landscape of possibilities for understanding beauty over time. As advancements in genetic research continue to emerge, the potential for harnessing gene editing for anti-aging purposes grows. By embracing healthy lifestyle choices and remaining informed about the genetic aspects of aging, individuals can navigate the aging process with confidence, enhancing their attractiveness and allure throughout their lives. As we continue to explore the implications of genetics in beauty, we find that aging can be seen not just as a decline in appearance, but as a natural part of a life journey that can still be celebrated and enhanced through informed choices and scientific advancements.

Chapter 13: Enhancing Features through Genetic Modification

In the evolving landscape of beauty and aesthetics, genetic modification presents a revolutionary frontier for enhancing physical attributes. This chapter explores the specific traits that can be targeted for enhancement through gene editing, the potential benefits of such modifications, and real-world examples that illustrate the implications of altering our genetic makeup for cosmetic purposes.

Specific Traits Targeted for Enhancement

Genetic modification has the potential to refine and enhance a variety of physical traits. Below are several areas where gene editing could play a significant role in beauty enhancement:

1. **Facial Symmetry**: Studies have shown that facial symmetry is often associated with attractiveness. Genetic editing could potentially target genes responsible for the growth and development of facial features, enabling modifications that create a more symmetrical appearance.

2. **Skin Quality**: The appearance of skin—its texture, elasticity, and pigmentation—can be modified through gene editing. By targeting genes involved in collagen production and skin cell regeneration, it may be possible to enhance skin quality, reduce wrinkles, and promote an even skin tone.

3. **Hair Density and Texture**: Genetic factors significantly influence hair characteristics such as density, curl pattern, and color. Gene editing could provide methods for individuals to modify these traits, leading to thicker hair or a preferred texture that aligns with current beauty trends.

4. **Body Shape and Size**: Although more controversial, gene editing could theoretically influence body composition, including fat distribution and muscle growth. This raises questions about the ethics of body modification, but the potential for such enhancements exists within the scientific community.

5. **Eye Color and Features**: Advances in gene editing could even allow for alterations in eye color or the enhancement of features such as eyelashes and eyebrows, leading to personalized aesthetic outcomes.

The Potential of Gene Editing to Alter Traits

The use of gene editing technologies, particularly CRISPR-Cas9, has opened the door to precise modifications that can influence beauty traits. CRISPR allows scientists to alter specific sequences of DNA, enabling the addition, removal, or alteration of genetic material at particular sites within the genome. This precision makes it a powerful tool for targeted enhancements:

- **Adding New Genetic Material**: By inserting genes associated with desirable traits (such as those promoting skin elasticity), individuals may enhance specific features.
- **Knocking Out Undesirable Genes**: In some cases, removing or disabling genes that lead to undesirable traits—such as excessive hair loss or premature aging—could improve overall appearance.

Examples of Successful Modifications and Their Implications

While much of the research in genetic modification is still in experimental stages, there have been promising developments in related fields:

1. **Cosmetic Gene Therapies**: Some companies are exploring the use of gene therapy to treat skin conditions like vitiligo or psoriasis. These treatments not only restore skin appearance but also improve overall skin health, serving as a form of beauty enhancement.

2. **Animal Models**: Research has demonstrated the use of CRISPR in animals to modify traits that enhance their aesthetic appeal—such as changing fur color in pets. These advancements provide insights into how similar techniques might one day be applied in humans.

3. **Ethical Considerations in Cosmetic Modifications**: While the technology shows promise, it also raises ethical questions. The potential for misuse, societal pressure to conform to specific beauty standards, and the implications of "designer" appearances must be carefully considered as we navigate this new landscape.

The Future of Genetic Modification in Beauty

As research in genetic editing continues to progress, the possibilities for enhancing beauty through genetics become increasingly tangible. The beauty industry is likely to see the emergence of personalized cosmetic solutions tailored to individual genetic profiles, leading to treatments that are as unique as the individuals seeking them.

However, with this potential comes the responsibility to approach genetic modification ethically and thoughtfully. Public discourse, informed consent, and regulatory frameworks will be essential in ensuring that these technologies are used to enhance, rather than harm, our perceptions of beauty and self-worth.

Conclusion

Genetic modification offers an exciting avenue for enhancing physical appearance, targeting traits that have long been sought after in the pursuit of beauty. As we stand on the brink of a new era in cosmetic enhancement, it is crucial to balance innovation with ethical considerations, ensuring that advancements in genetic editing serve to empower individuals rather than impose narrow definitions of attractiveness. Through responsible exploration of these technologies, we can harness the power of genetics to unlock our fullest potential, both inside and out.

Chapter 14: Holistic Approaches to Beauty

In our pursuit of beauty, it is essential to consider a comprehensive perspective that integrates both genetic and non-genetic factors. Holistic beauty recognizes that attractiveness is not solely derived from genetic enhancements or modifications, but also encompasses lifestyle choices, mental and emotional well-being, and the incorporation of traditional beauty practices. This chapter explores how a holistic approach to beauty can enhance our overall appearance and well-being, combining advancements in genetic science with time-honored techniques.

Integrating Genetic and Non–Genetic Methods for Enhancing Beauty

The Power of Personalization

- Genetic modification offers personalized solutions to enhance physical traits. However, integrating these with non-genetic methods can maximize results. For instance, someone may opt for gene editing to improve skin elasticity, while also adopting a skincare regimen tailored to their genetic profile.
- Genetic testing can reveal predispositions to certain skin conditions or sensitivities, allowing for personalized skincare products and routines that complement genetic enhancements.

Mental and Emotional Well-Being

- Attractiveness is often linked to how we feel about ourselves. Psychological research suggests that individuals who perceive themselves as beautiful are more likely to engage socially and confidently.

- Techniques such as mindfulness, meditation, and positive affirmations can help enhance self-esteem and body image. When combined with genetic enhancements, these practices can create a robust framework for feeling and looking beautiful.

Traditional Beauty Practices

- Many cultures have rich traditions of beauty practices that predate modern science. For instance, herbal remedies, essential oils, and natural skincare treatments have been used for centuries to maintain skin health and radiance.

- Integrating these traditional practices with modern genetic insights can provide a multifaceted approach to beauty. For example, using gene editing to enhance skin health could be paired with herbal treatments known for their rejuvenating properties.

The Role of Lifestyle Choices in Enhancing Beauty

Nutrition

- As discussed in earlier chapters, diet plays a crucial role in gene expression and overall health. Nutritional choices can influence everything from skin texture to hair quality.
- A holistic beauty approach emphasizes the consumption of nutrient-rich foods that support genetic health. This includes foods high in antioxidants, vitamins, and omega-3 fatty acids, which can promote a vibrant and youthful appearance.

Physical Activity

- Regular exercise not only enhances physical fitness but also contributes to mental well-being. Physical activity can improve circulation, promote healthy skin, and enhance muscle tone—all vital components of physical attractiveness.
- Integrating exercise routines that align with personal goals and genetic profiles can further amplify results. For example, someone with a genetic predisposition to certain muscular developments may tailor their workout regimen to capitalize on those strengths.

Stress Management

- Chronic stress can have detrimental effects on physical appearance, including premature aging and skin issues. A holistic beauty regimen incorporates stress management techniques such as yoga, meditation, and hobbies that promote relaxation and joy.
- Managing stress effectively can enhance the results of genetic modifications and contribute to a more attractive appearance by fostering a calm, confident demeanor.

Combining Traditional Beauty Practices with Genetic Advancements

Natural Ingredients

- Many traditional beauty practices utilize natural ingredients that can work synergistically with genetic modifications. For instance, oils like argan or coconut can provide moisture and nutrients that support skin health enhanced by gene editing.
- The synergy between genetic advancements and natural ingredients allows for more comprehensive beauty solutions that address both the surface and the underlying genetic predispositions.

Cultural Wisdom

- Different cultures offer unique perspectives on beauty that emphasize balance and well-being. For example, traditional Asian beauty routines often focus on holistic practices that include diet, skincare, and mindfulness.
- By incorporating cultural wisdom into modern beauty routines, individuals can create a personalized approach that respects both their genetic makeup and cultural heritage.

Conclusion: A Holistic View of Attractiveness

As we advance into a future where genetic enhancements are increasingly accessible, it is essential to maintain a holistic perspective on beauty. The integration of genetic and non-genetic methods provides a comprehensive approach that addresses the physical, mental, and emotional aspects of attractiveness. By embracing lifestyle choices, traditional practices, and the insights of genetic science, individuals can cultivate a beauty that is not only appealing but also deeply rooted in well-being and self-acceptance.

In the chapters to come, we will further explore how societal standards of beauty are influenced by these holistic approaches and the potential innovations that lie ahead in the realm of genetic beauty.

Chapter 15: Societal Standards of Beauty

Beauty is a concept that transcends the individual, woven into the fabric of society. Its definitions are fluid, shaped by historical context, cultural narratives, and, increasingly, scientific advancements. In this chapter, we will explore the evolution of beauty standards, the role of media and technology in shaping perceptions, and the potential for genetic advancements to challenge and redefine societal norms around beauty.

Historical and Cultural Contexts of Beauty Standards

Beauty Through the Ages

- Throughout history, notions of beauty have varied dramatically across cultures and eras. From the curvaceous figures celebrated in the Renaissance to the lean silhouettes of the modern age, societal standards reflect changing values and ideals.
- In Ancient Egypt, beauty was synonymous with symmetry and clear skin, while in the Victorian era, a fuller figure was often considered more desirable, reflecting wealth and prosperity.

Cultural Influences

- Different cultures have their own unique standards of beauty. For example, in some African cultures, body modifications such as scarification or lip plates are viewed as beautiful, while in Western cultures, ideals often revolve around skin tone, hair texture, and body size.
- The globalization of media has introduced diverse beauty ideals, sometimes creating tensions between traditional values and contemporary trends.

The Influence of Media and Technology

Media Representations

- The portrayal of beauty in media plays a pivotal role in shaping public perception. Advertising, film, and social media can perpetuate narrow standards, often favoring specific body types, skin tones, and facial features.
- The rise of social media influencers has introduced a new dynamic, where beauty is often commodified, leading to a culture of comparison and aspiration. This can have positive or negative effects on self-esteem and body image.

The Role of Technology

- Advances in technology have not only changed how beauty is represented but have also influenced how individuals pursue beauty. Filters, editing software, and beauty apps can create unrealistic standards, leading to a disparity between appearance and reality.
- Conversely, technology has also empowered individuals to challenge these norms. Social media platforms can amplify voices advocating for body positivity and diversity, encouraging broader acceptance of various beauty ideals.

Challenging Societal Norms Through Genetic Advancements

Redefining Beauty with Genetic Insights

- The potential of genetic modification to enhance physical traits opens up conversations about what beauty can mean. As gene editing technologies become more accessible, individuals may choose to modify features traditionally deemed unattractive or undesirable.
- By democratizing beauty enhancement, genetic advancements may reduce the pressure to conform to societal standards that often marginalize certain groups based on their genetic traits.

The Impact of Inclusivity

- The integration of genetic advancements could promote inclusivity, enabling a diverse range of beauty to be celebrated. By allowing individuals to express themselves genetically, we may see a shift towards a broader understanding of beauty that encompasses various shapes, sizes, and features.
- This inclusivity can challenge the prevailing stereotypes that often dictate who is deemed attractive, encouraging acceptance of diverse identities.

Conclusion: The Future of Beauty Standards

As we navigate a world increasingly influenced by genetic advancements, the standards of beauty will continue to evolve. With the power of gene editing at our fingertips, we have the opportunity to redefine attractiveness in a way that values individuality over conformity.

The challenge lies not only in harnessing genetic science for enhancement but also in fostering a culture that embraces diversity and self-acceptance. By critically engaging with societal standards and advocating for broader definitions of beauty, we can pave the way for a future where everyone has the opportunity to feel attractive and valued, regardless of their genetic makeup.

In the next chapter, we will explore the emerging innovations in genetic beauty, examining how advancements in gene editing may shape the landscape of beauty treatments and personal care in the years to come.

Chapter 16: Future Innovations in Genetic Beauty

As we stand on the cusp of a new era in beauty and aesthetics, the interplay between genetics and technology offers exciting possibilities. With rapid advancements in gene editing technologies, particularly CRISPR and other innovative approaches, the beauty industry is poised for transformation. This chapter will explore emerging trends in genetic beauty, predictions for future applications, and potential breakthroughs that could redefine how we perceive and achieve attractiveness.

Emerging Trends in Gene Editing and Cosmetic Applications

Personalized Beauty Solutions

- The future of beauty is increasingly leaning towards personalization. Genetic testing allows individuals to identify their unique genetic makeup, enabling tailored beauty solutions that address specific needs and preferences. From skin care formulations to dietary recommendations, personalization will drive new product development.
- Brands are beginning to embrace genetic insights in their marketing strategies, offering customized beauty products based on individual genetic profiles. This trend is expected to grow as more consumers seek products that align with their biological needs.

Innovations in Cosmetic Gene Editing

- The application of gene editing to enhance physical features is no longer a distant dream. Researchers are exploring the feasibility of using gene editing technologies to target specific traits, such as improving skin elasticity, enhancing hair thickness, or even modifying pigmentation.

- Companies are experimenting with topical applications of gene-editing technologies that could allow consumers to enhance their features at home. This innovation could revolutionize the way beauty treatments are administered, making advanced aesthetics accessible to a broader audience.

Integration of Artificial Intelligence

- AI is increasingly being integrated into genetic beauty applications, allowing for advanced data analysis and predictive modeling. AI algorithms can analyze genetic data alongside lifestyle factors, providing personalized recommendations for beauty treatments and interventions.

- Virtual consultations with geneticists and beauty professionals facilitated by AI technologies will likely become standard practice, ensuring that consumers receive tailored advice based on their unique genetic backgrounds.

Predictions for the Future of Beauty Through Genetic Modification

Mainstream Acceptance of Genetic Beauty

- As understanding and acceptance of genetic modifications grow, the stigma surrounding genetic enhancements is expected to diminish. This shift will pave the way for more individuals to explore gene editing as a viable option for improving their appearance.
- Educational initiatives will play a crucial role in fostering acceptance, as consumers become more informed about the safety, efficacy, and ethical implications of genetic enhancements.

Advancements in Anti-Aging Treatments

- Research into gene therapies aimed at reversing or slowing the aging process is rapidly advancing. Innovations in this area could lead to treatments that not only enhance appearance but also improve overall skin health and vitality.
- Future innovations may include gene editing approaches that promote the regeneration of collagen, elastin, and other critical components of youthful skin, providing effective solutions for aging-related changes.

Customized Nutritional Solutions

- The integration of nutrigenomics into beauty regimens will continue to grow. Personalized diets that align with genetic profiles will enhance beauty from the inside out, optimizing skin health, hair quality, and overall appearance.
- As consumers seek holistic approaches to beauty, the emphasis on nutrition in genetic beauty routines will drive demand for supplements and food products designed to support genetic expression.

Potential Breakthroughs on the Horizon

Gene Editing in Cosmetic Surgery

- The future may see gene editing techniques being applied in conjunction with cosmetic surgery. For instance, pre-surgical genetic modifications could enhance the healing process or improve aesthetic outcomes, resulting in more satisfying results for patients.
- Surgeons could utilize genetic insights to tailor procedures to an individual's specific genetic predispositions, minimizing risks and optimizing recovery.

Regenerative Medicine Applications

- Advances in regenerative medicine, combined with gene editing, could lead to groundbreaking treatments for scarring, pigmentation issues, and other beauty-related concerns. Techniques such as stem cell therapy might be enhanced through genetic modifications, providing innovative solutions for skin rejuvenation.

- Future research may focus on using gene editing to promote the body's natural healing processes, leading to effective treatments that enhance beauty without invasive procedures.

Global Accessibility of Genetic Beauty Solutions

- As technology becomes more affordable and accessible, genetic beauty solutions may be democratized, allowing individuals from diverse backgrounds to explore enhancements previously limited to the privileged few.

- Efforts to ensure equitable access to genetic beauty innovations will be crucial in shaping a future where everyone can benefit from advancements in the field.

Conclusion

The future of genetic beauty is filled with promise and potential. As we harness the power of gene editing, the possibilities for enhancing physical attractiveness are boundless. By integrating genetic insights into beauty practices, we can redefine beauty standards and create a more inclusive, personalized approach to aesthetics.

In the next chapter, we will delve into the legal and regulatory considerations surrounding genetic editing, examining the guidelines that will shape the future of this rapidly evolving field. Understanding these frameworks will be essential for navigating the complexities of genetic beauty advancements responsibly.

Chapter 17: Legal and Regulatory Considerations

As the field of genetic editing and modification advances, it is imperative to address the legal and regulatory frameworks that will govern these innovations, particularly in the context of beauty enhancements. This chapter will provide an overview of the current laws related to genetic editing, discuss ethical guidelines in cosmetic genetic research, and explore potential future legal challenges within the beauty industry.

Current Laws Governing Genetic Editing and Modification

Regulatory Bodies and Frameworks

- In the United States, the Food and Drug Administration (FDA) is the primary regulatory body overseeing gene editing technologies, particularly when they are applied to human subjects. The FDA ensures that any products or procedures utilizing gene editing meet safety and efficacy standards before they can be marketed to consumers.
- In Europe, the European Medicines Agency (EMA) provides similar oversight, and gene therapies must undergo rigorous clinical trials and evaluations before receiving approval for use. Additionally, the European Union has stringent regulations concerning genetically modified organisms (GMOs) that impact gene editing research and application.

Legislation on Gene Editing

- Various countries have enacted specific laws regulating gene editing practices. For example, the UK has developed guidelines that differentiate between somatic gene editing (which targets non-reproductive cells) and germline editing (which affects reproductive cells and can be passed on to future generations). While somatic editing is largely permitted, germline editing is subject to stricter regulations due to ethical concerns.
- In some regions, laws explicitly prohibit any genetic modifications aimed at enhancing human traits for non-therapeutic purposes, thus limiting the scope of genetic beauty applications.

Informed Consent

Ethical considerations mandate that individuals undergoing gene editing procedures must provide informed consent. This includes a comprehensive understanding of the potential risks, benefits, and long-term implications of genetic modifications, especially when the aim is cosmetic enhancement.

Ethical Guidelines for Cosmetic Genetic Research

Guidelines for Research

- Ethical guidelines developed by various professional organizations, such as the American Society of Gene & Cell Therapy (ASGCT), emphasize the need for responsible research practices. These guidelines advocate for the fair treatment of research subjects, transparent reporting of findings, and ongoing monitoring of long-term effects post-treatment.
- Researchers are encouraged to engage in public discourse about the implications of their work, fostering a collaborative dialogue that includes ethicists, policymakers, and community representatives.

Risk Assessment

Researchers must conduct thorough risk assessments before initiating studies on gene editing for cosmetic purposes. This involves evaluating the potential for unintended consequences, both on an individual level and within the broader societal context. Ethical frameworks stress the importance of minimizing risks and ensuring that potential benefits justify any associated harms.

Equity and Access

Ethical guidelines also address concerns regarding equity and access to genetic enhancements. As genetic beauty solutions emerge, there is a risk that only certain populations may benefit from these advancements, exacerbating existing inequalities in beauty standards and access to care. Addressing these disparities is a critical ethical consideration in the development of genetic beauty solutions.

Future Legal Challenges and Considerations in the Beauty Industry

Intellectual Property Issues

As genetic beauty products enter the market, intellectual property rights will become increasingly complex. Patent laws will need to evolve to address the ownership of genetic modifications, especially when they involve naturally occurring genetic sequences. This may lead to legal disputes regarding the rights to use certain genetic edits for cosmetic purposes.

Global Disparities in Regulation

The international nature of the beauty market presents challenges in harmonizing regulations across borders. Different countries may have varying stances on the permissibility of gene editing for cosmetic purposes, leading to potential legal conflicts for multinational beauty companies. Global standards may need to be developed to ensure safety and ethical compliance.

Emerging Technologies and Ethical Dilemmas

With the rapid pace of innovation in genetic editing technologies, new ethical dilemmas will arise. As techniques evolve, the potential for misuse or unintended consequences could lead to public backlash and legal challenges. Ongoing dialogue among scientists, ethicists, and lawmakers will be essential to navigate these emerging issues responsibly.

Public Perception and Acceptance

The acceptance of genetic enhancements in beauty will also influence legal frameworks. Public perception can drive regulatory changes, as societal attitudes shift towards or against genetic modification. Engaging with the community and fostering transparency will be vital in shaping a legal landscape that reflects public values.

Conclusion

The legal and regulatory considerations surrounding genetic editing in beauty are complex and evolving. As innovations in gene editing technologies continue to unfold, it is essential to establish a framework that prioritizes safety, ethical integrity, and equitable access. By addressing the challenges and opportunities presented by genetic beauty enhancements, we can work towards a future that harmonizes scientific advancement with ethical responsibility.

In the next chapter, we will explore real-life applications of genetic editing, examining case studies of individuals who have utilized these technologies for beauty enhancements and the impacts on their lives.

Chapter 18: Real-Life Applications of Genetic Editing

As the field of genetic editing continues to evolve, its applications are becoming increasingly prominent in the realm of beauty. This chapter explores real-life case studies of individuals who have utilized genetic modification to enhance their appearance, highlighting both success stories and the challenges faced. Additionally, we will examine the broader implications of these transformations on perceptions of beauty in society.

Case Studies of Genetic Modification for Beauty

1. The Transformation of Emily: A Case of Hair Thickness

Emily, a 29-year-old woman with a family history of thinning hair, decided to undergo a genetic modification procedure aimed at enhancing the thickness and fullness of her hair. After a genetic test revealed specific variants associated with hair growth, she opted for a targeted treatment that involved editing genes responsible for the regulation of hair follicles.

Outcome: After a six-month treatment regimen, Emily experienced significant improvements in hair thickness and volume. This transformation not only boosted her confidence but also changed her social interactions. Emily reported feeling more attractive and receiving compliments that positively affected her self-esteem.

Challenges: While Emily's case was largely successful, she encountered some side effects during the initial phase of treatment, including minor scalp irritation. These challenges highlighted the importance of thorough consultations and monitoring during genetic procedures.

2. David's Journey: From Scarring to Smooth Skin

David, a 35-year-old man who had endured severe acne scarring during his teenage years, sought genetic editing to improve his skin texture. After consultations with dermatologists and geneticists, David participated in a clinical trial utilizing CRISPR technology to edit genes involved in skin healing and collagen production.

Outcome: Over the course of a year, David saw remarkable changes. The scarring on his face diminished significantly, and he developed smoother skin. The transformation not only enhanced his physical appearance but also provided him with a newfound confidence to engage socially and professionally.

Challenges: Despite the positive results, David faced emotional challenges during the recovery period, particularly due to the visible effects of the treatment before they improved. The psychological burden of waiting for results taught him the importance of patience and realistic expectations when undergoing genetic modifications.

3. Sarah's Experience: Enhancing Facial Symmetry

Sarah, an aspiring model, approached genetic enhancement to improve facial symmetry, a trait often associated with beauty. After extensive genetic testing, she opted for a procedure that included precise editing of genes related to facial structure.

Outcome: Post-procedure, Sarah noticed a subtle yet impactful enhancement in her facial symmetry. This change positively influenced her modeling career, as she began to receive more job offers and gained visibility in the industry.

Challenges: Sarah's journey included navigating societal perceptions of beauty. While many praised her new appearance, she also faced criticism for altering her natural look. This experience underscored the societal complexities surrounding beauty enhancements and the pressure to conform to certain standards.

The Impact on Lives and Perceptions of Beauty
Personal Transformation

The individuals featured in these case studies experienced profound changes, not just in their physical appearance but also in their self-esteem, social interactions, and professional opportunities. Genetic enhancements allowed them to address specific beauty concerns, leading to enhanced confidence and quality of life.

Societal Implications

The success stories of Emily, David, and Sarah reflect a broader trend in society where genetic editing is becoming more accepted as a means of achieving beauty. However, these advancements also raise ethical questions and societal dilemmas. As genetic enhancements become more common, discussions surrounding authenticity, societal norms, and the definition of beauty are increasingly relevant.

Conclusion

Real-life applications of genetic editing for beauty illustrate the transformative potential of this technology. While the benefits are evident, the experiences of individuals also highlight the complexities and challenges that accompany such modifications. As the beauty industry continues to embrace genetic advancements, it is essential to maintain a dialogue about the implications of these changes and to foster an inclusive understanding of beauty that transcends mere physical appearance.

In the next chapter, we will delve into practical strategies for building a genetic beauty routine, providing actionable steps for incorporating genetic knowledge into everyday beauty practices.

Chapter 19: Building a Genetic Beauty Routine

As genetic advancements continue to reshape the beauty landscape, integrating this knowledge into daily routines can significantly enhance personal aesthetics and overall well-being. This chapter focuses on practical steps for creating a genetic beauty routine, including actionable strategies, recommended products, and techniques that leverage genetic insights. By understanding how our genetics interact with beauty practices, individuals can develop personalized plans that align with their unique genetic profiles.

Understanding Your Genetic Makeup

1. Genetic Testing

Before embarking on a genetic beauty routine, the first step is to understand your genetic makeup. Genetic testing can provide valuable insights into traits that affect appearance, such as skin type, hair texture, and predispositions to certain conditions.

- **Choosing a Testing Service**: Various companies offer genetic testing focused on beauty and health. Look for reputable providers that offer comprehensive reports on genetic traits relevant to beauty.

- **Interpreting Results**: Engage with a genetic counselor or specialist to help interpret your results. This step is crucial for understanding how your genetics influence your beauty regimen.

Practical Steps for Incorporating Genetic Knowledge

2. Tailoring Your Skincare Routine

Based on your genetic insights, you can customize your skincare routine to target specific concerns effectively.

- **Identifying Skin Type**: Genetic testing can reveal your skin's predisposition to conditions like acne, dryness, or sensitivity. Choose products formulated for your skin type.

- **Incorporating Active Ingredients**: Depending on genetic factors, certain active ingredients may work better for you. For example:

- **If predisposed to hyperpigmentation**, look for products containing vitamin C or niacinamide.

- **For sensitive skin**, consider calming ingredients like chamomile or aloe vera.

3. Hair Care Customized for Your Genetic Profile

Your hair's texture, growth patterns, and health can also be influenced by genetics. Use this knowledge to create a personalized hair care routine.

- **Understanding Hair Type**: Know whether your hair is straight, wavy, curly, or coily, and choose shampoos and conditioners that cater to your specific type.
- **Addressing Hair Concerns**: If your genetic profile indicates a predisposition to thinning hair, consider incorporating treatments like minoxidil or supplements that promote hair health, such as biotin and omega-3 fatty acids.

4. Nutrition for Beauty

Nutrition plays a critical role in enhancing beauty from the inside out. Nutrigenomics, the study of how food interacts with genes, can guide dietary choices to promote an attractive appearance.

Food Choices

- **If predisposed to skin aging**, increase your intake of antioxidants through berries, leafy greens, and nuts.
- **For hair health**, prioritize foods rich in protein, iron, and zinc, such as lean meats, legumes, and whole grains.

5. Exercise and Lifestyle Modifications

Incorporating exercise and healthy lifestyle choices can optimize genetic expression and improve physical appearance.

- **Regular Physical Activity**: Aim for at least 150 minutes of moderate exercise each week. Activities like strength training can enhance muscle tone and improve body composition.
- **Stress Management**: Chronic stress can negatively impact both physical appearance and overall health. Practices like meditation, yoga, or mindfulness can reduce stress and improve mental well-being.

Recommended Products and Techniques
6. Product Selection

Choose products that complement your genetic profile and address specific beauty concerns. Here are some categories to consider:

Skincare

- Sunscreens with high SPF and broad-spectrum protection.
- Moisturizers enriched with ingredients suited to your skin type.

Haircare

- Sulfate-free shampoos and conditioners tailored for your hair type.
- Leave-in treatments that target specific concerns (e.g., frizz control, moisture retention).

7. Techniques to Enhance Beauty

In addition to products, consider adopting techniques that align with your genetic profile:

- **Facial Massage**: This can improve circulation and promote lymphatic drainage, leading to a more radiant complexion.
- **Personalized Makeup Application**: Learn techniques that highlight your best features based on your facial structure and skin tone.

Building a Personalized Beauty Plan

8. Creating Your Beauty Routine

1. **Assess Your Genetic Profile**: Start by reviewing your genetic testing results.
2. **Define Your Goals**: Identify specific beauty goals you wish to achieve (e.g., clearer skin, thicker hair).
3. **Select Products and Practices**: Choose products and techniques that align with your genetics and goals.
4. **Monitor Progress**: Keep track of your beauty routine's effectiveness and adjust as needed.

9. Consulting with Professionals

Consider consulting with dermatologists, nutritionists, or beauty professionals who can provide tailored advice based on your genetic makeup and beauty aspirations. Their expertise can help refine your routine for optimal results.

Conclusion

Building a genetic beauty routine offers a personalized approach to enhancing your appearance while taking advantage of the latest advancements in genetics. By understanding your unique genetic traits and making informed choices about skincare, hair care, nutrition, and lifestyle, you can harness your genetic potential for an attractive and alluring presence. As we move forward, the next chapter will explore community perspectives on genetic beauty, emphasizing the importance of shared experiences and support in this evolving field.

Chapter 20: Community Perspectives on Genetic Beauty

As we explore the evolving landscape of beauty shaped by genetic advancements, it becomes evident that community plays a pivotal role in how individuals perceive and embrace genetic beauty practices. This chapter delves into the importance of community support, discussions surrounding genetic enhancement, and the creation of networks for sharing experiences and insights.

The Importance of Community Support

1. Building a Supportive Environment

The journey of embracing genetic beauty often comes with uncertainties and societal pressures. Having a supportive community can significantly influence an individual's confidence and acceptance of their choices.

- **Encouragement and Validation**: Community support provides encouragement during the often challenging process of exploring genetic modifications. Sharing personal experiences can validate feelings and choices, making individuals feel less isolated in their journeys.

- **Sharing Success Stories**: Communities can foster positivity by sharing success stories of individuals who have benefitted from genetic advancements. These narratives can inspire others and provide practical insights into the transformative potential of genetic beauty practices.

2. Reducing Stigmas

Genetic modification for beauty purposes can provoke mixed reactions from society. Community discussions can help to normalize these practices and reduce the stigma associated with them.

- **Education and Awareness**: Engaging in open dialogues within communities helps educate members about the science of genetics and the benefits of gene editing. Addressing misconceptions can pave the way for broader acceptance.
- **Creating Safe Spaces**: Establishing forums and support groups can offer safe environments for individuals to express concerns, share experiences, and seek guidance without judgment.

Forums and Networks for Sharing Experiences
3. Online and Offline Communities

The rise of digital technology has facilitated the formation of online communities where individuals interested in genetic beauty can connect.

- **Social Media Platforms**: Platforms like Facebook, Instagram, and specialized forums can be invaluable resources for individuals seeking information, inspiration, or camaraderie. They allow members to share tips, experiences, and recommendations related to genetic beauty.
- **In-Person Events**: Workshops, seminars, and conferences focusing on genetic advancements in beauty can provide networking opportunities and foster deeper connections among participants. Such events can feature guest speakers, expert panels, and discussions that address both the scientific and emotional aspects of genetic beauty.

4. Collaborative Learning

Communities centered around genetic beauty can serve as hubs for collaborative learning.

- **Knowledge Sharing**: Individuals can exchange valuable information about genetic testing, product recommendations, and effective techniques for enhancing beauty. This collaborative approach empowers members to make informed decisions about their beauty regimens.

- **Expert Contributions**: Involving geneticists, dermatologists, and nutritionists in community discussions can further enrich the knowledge pool. Their expertise can help demystify complex concepts and provide professional insights into genetic beauty practices.

Building a Community Around Genetic Beauty

5. Creating Inclusive Spaces

To build a thriving community around genetic beauty, inclusivity is paramount.

- **Diverse Perspectives**: Embracing diverse backgrounds and experiences can enrich discussions. Individuals from various cultural, racial, and socio-economic backgrounds can offer unique perspectives on beauty standards and genetic practices.

- **Fostering Respect and Understanding**: It's essential to cultivate an environment of respect and understanding, where differing opinions about genetic modifications are acknowledged and explored thoughtfully.

6. Advocacy and Activism

Communities can also play an active role in advocating for responsible genetic practices.

- **Raising Awareness**: Communities can engage in campaigns that promote awareness about the ethical use of gene editing in beauty, ensuring that discussions are informed by current research and ethical guidelines.

- **Policy Influence**: By collectively voicing concerns and suggestions, communities can influence policies related to genetic editing and cosmetic applications, contributing to a more responsible approach to genetic enhancements.

Conclusion

The significance of community in the realm of genetic beauty cannot be overstated. By fostering supportive environments, encouraging open dialogue, and providing platforms for sharing experiences, communities can empower individuals to embrace genetic advancements confidently. As we move into the next chapter, we will explore the dimensions of beauty beyond genetics, emphasizing how personality, talent, and intelligence contribute to a holistic understanding of attractiveness.

Chapter 21: Beauty Beyond Genetics

While genetic advancements offer exciting possibilities for enhancing physical appearance, it is crucial to recognize that beauty is a multifaceted concept that extends beyond mere genetic traits. This chapter delves into the various dimensions of beauty, including personality, talent, intelligence, and how these attributes can complement genetic beauty. Emphasizing a holistic approach to attractiveness allows us to appreciate the full spectrum of what makes an individual truly alluring.

Exploring Other Dimensions of Beauty

1. Personality: The Inner Glow

Personality plays a pivotal role in defining attractiveness. Attributes such as kindness, humor, confidence, and charisma can enhance one's appeal significantly.

- **Charisma and Attractiveness**: Charisma is often seen as an essential trait that draws people in. Charismatic individuals tend to exude confidence and warmth, making them more attractive to others regardless of their physical appearance.
- **Positive Attitude**: A positive attitude can enhance attractiveness, as people are naturally drawn to those who radiate happiness and optimism. This aspect of personality can often overshadow physical traits, creating deeper connections with others.

2. Talent: The Beauty of Skills

Talents and skills can add an alluring dimension to an individual's personality. Whether it's artistic abilities, athletic prowess, or intellectual achievements, talents can significantly contribute to one's attractiveness.

- **Creativity and Expression**: Artistic talents—such as music, painting, or writing—can showcase creativity and emotional depth, which many find highly appealing. A person who can express themselves through their art can evoke admiration and respect, enhancing their overall allure.
- **Intellectual Abilities**: Intelligence and critical thinking skills can also enhance attractiveness. Engaging conversations and a deep understanding of various topics can create a magnetic pull, leading to connections based on shared interests and mutual respect.

3. Intelligence: The Allure of the Mind

Intelligence is not merely a measure of academic success; it encompasses emotional intelligence, social awareness, and the ability to navigate complex situations.

- **Emotional Intelligence**: Understanding and managing emotions, both one's own and those of others, can foster deeper relationships. People with high emotional intelligence tend to be more empathetic and communicative, which can enhance their attractiveness significantly.
- **Conversational Skills**: The ability to engage in stimulating conversations can captivate others. Intelligent discourse, combined with a sense of humor and insight, can leave a lasting impression, making someone memorable and attractive.

Complementing Genetic Beauty
4. The Harmony of Attributes

Genetic beauty can serve as a foundation, but when combined with strong personality traits, talents, and intelligence, it creates a well-rounded individual whose attractiveness is multidimensional.

- **Holistic Attractiveness**: Individuals who possess a balance of physical, emotional, and intellectual traits are often perceived as more attractive. The interplay between genetics and other qualities leads to a deeper connection with others, fostering relationships built on more than just physical appearances.
- **Cultural Influences**: Different cultures may emphasize various aspects of beauty. In some societies, inner qualities such as wisdom and compassion may be valued above physical traits, showcasing the diversity of beauty standards worldwide.

Embracing a Holistic View of Attractiveness
5. Moving Beyond Superficiality

To cultivate a genuine sense of beauty, it's essential to move beyond superficial judgments based solely on genetic traits. Embracing a holistic view encourages a deeper understanding of what makes someone truly attractive.

- **Self-Acceptance**: Encouraging individuals to embrace all facets of their identity, including their physical attributes, personality, talents, and intelligence, promotes self-acceptance and confidence. This mindset can lead to more fulfilling relationships and interactions.
- **Community and Connection**: Building communities that celebrate diverse forms of beauty—embracing personality, talent, and intelligence—can help foster acceptance and appreciation for different attributes. Discussions around beauty should encompass all dimensions, allowing individuals to feel valued for who they are as a whole.

Conclusion

As we continue to explore the realm of genetic beauty, it is vital to remember that true attractiveness lies in the synergy of genetics and the myriad of personal attributes that shape an individual. By celebrating the diversity of beauty and acknowledging the importance of personality, talent, and intelligence, we pave the way for a more inclusive and holistic understanding of what it means to be truly attractive. In the subsequent chapter, we will address the stigmas associated with genetic modification, working toward a more informed and accepting perspective on this evolving field.

Chapter 22: Overcoming Stigmas Associated with Genetic Modification

As the field of genetic modification and enhancement continues to advance, it faces significant societal stigma rooted in misconceptions, fears, and ethical concerns. This chapter aims to address these stigmas, promote understanding, and foster acceptance of genetic beauty practices. By educating individuals and communities about the realities of genetic editing and its applications in beauty, we can create a more inclusive dialogue that embraces advancements in this transformative field.

Addressing Misconceptions and Fears Surrounding Gene Editing

1. Understanding the Technology

One of the primary barriers to acceptance of genetic editing is a lack of understanding of the technology itself. Many people conflate gene editing with more extreme forms of genetic manipulation, such as eugenics or cloning, leading to fear and resistance.

- **Educating the Public**: To combat these fears, educational initiatives should aim to demystify genetic editing technologies like CRISPR. Public lectures, workshops, and easily accessible resources can help clarify what gene editing entails, how it works, and its potential benefits.

- **Success Stories**: Highlighting successful applications of gene editing in medicine, agriculture, and cosmetics can illustrate its positive impacts and the ethical frameworks guiding its use.

2. The Ethical Debate

The ethical implications of genetic modification often dominate discussions surrounding its acceptance. Concerns about "playing God," potential health risks, and social inequality need to be addressed thoughtfully.

- **Ethical Frameworks**: Engaging ethicists, scientists, and community leaders in open dialogues can help establish ethical frameworks that govern genetic modifications. These discussions should involve diverse perspectives to ensure a comprehensive understanding of societal values.
- **Regulation and Oversight**: Clear regulations and guidelines for genetic editing practices can help alleviate fears by ensuring that modifications are safe, effective, and ethically conducted.

Fostering Acceptance of Genetic Beauty Practices
3. Promoting Inclusivity

Acceptance of genetic beauty practices is closely tied to broader societal attitudes toward diversity and inclusivity. By framing genetic enhancements as tools for empowerment rather than elitism, we can foster a more positive perception.

- **Diverse Representation**: Showcasing individuals from varied backgrounds who utilize genetic beauty enhancements can promote inclusivity. Media campaigns, social media stories, and public figures advocating for responsible gene editing can help normalize the practice.

- **Community Involvement**: Encouraging communities to engage in discussions about genetic beauty can help build acceptance. Workshops, forums, and community events can provide platforms for sharing experiences and learning from one another.

4. Education as a Means to Reduce Stigma

Education is key to dispelling myths and reducing stigma surrounding genetic modification. By providing accurate information, we can empower individuals to make informed decisions about their beauty practices.

- **Workshops and Seminars**: Organizing educational events focused on the science behind genetic editing and its applications in beauty can help demystify the process. Inviting experts to speak and engage with the community fosters a supportive environment for learning.

- **Digital Resources**: Creating online platforms with credible resources, articles, and FAQs about genetic beauty can help reach a wider audience. These resources can serve as go-to guides for anyone interested in learning more about the topic.

Building a Supportive Community

5. Creating Networks for Open Dialogue

Supportive communities play a vital role in overcoming stigma associated with genetic modifications. Building networks where individuals can share their experiences, ask questions, and provide support is essential.

- **Peer Support Groups**: Establishing support groups where individuals can discuss their experiences with genetic beauty practices can create a sense of community. These spaces can offer emotional support and a platform for exchanging information and resources.
- **Online Forums**: Developing online forums or social media groups focused on genetic beauty can facilitate discussions and create a sense of belonging. These platforms can also provide a space for individuals to share their journeys and successes.

6. Advocating for Acceptance

Advocacy plays a crucial role in shifting societal attitudes toward genetic beauty. By highlighting the benefits and ethical considerations, we can work to create a more accepting environment.

- **Campaigns and Initiatives**: Organizing campaigns that promote genetic beauty as a legitimate practice can help raise awareness and foster acceptance. Collaborating with organizations that advocate for scientific advancements can amplify these efforts.
- **Engaging Influencers**: Involving influencers and public figures in discussions about genetic beauty can help normalize the practice. Their platforms can reach a broad audience, sparking interest and dialogue.

Conclusion

Overcoming the stigmas associated with genetic modification requires a collective effort that includes education, open dialogue, and community support. By addressing misconceptions, fostering inclusivity, and advocating for responsible practices, we can pave the way for a more accepting attitude toward genetic beauty. As we move forward, the subsequent chapter will explore the balance between ethics and beauty, highlighting the importance of thoughtful engagement in the ever-evolving landscape of genetic advancements.

Chapter 23: Finding Balance: Ethics and Beauty

As genetic editing technologies continue to evolve, the intersection of beauty, ethics, and genetics presents both exciting opportunities and formidable challenges. This chapter will explore how to navigate this complex landscape, emphasizing the need for a balanced approach that honors individual desires for beauty while upholding ethical standards.

Navigating the Ethical Landscape of Genetic Enhancement

1. Defining Ethical Boundaries

The advancements in genetic editing raise essential questions about what is considered acceptable or ethical when it comes to modifying human traits for aesthetic purposes. It is crucial to establish clear boundaries to ensure that genetic enhancements are pursued responsibly.

- **Informed Consent**: A fundamental principle of ethics in any medical or scientific field is the requirement for informed consent. Individuals considering genetic enhancements must fully understand the implications, risks, and benefits of the procedures.
- **Risk-Benefit Analysis**: Before undertaking any form of genetic enhancement, a thorough assessment of potential risks versus expected benefits must be conducted. This includes evaluating not only physical health outcomes but also psychological and social implications.

2. Equity and Access

One of the significant ethical concerns surrounding genetic beauty practices is the potential for creating a divide between those who can afford enhancements and those who cannot. This disparity can lead to a society where beauty standards are further influenced by socioeconomic status.

- **Addressing Inequality**: Efforts must be made to ensure equitable access to genetic enhancements. This may involve policies that support lower-cost solutions, research funding for accessibility, and partnerships with organizations dedicated to health equity.
- **Public Health Initiatives**: Governments and health organizations should consider public health initiatives that educate individuals about the ethical implications of genetic enhancement and work to make these technologies accessible to all.

Balancing Desire for Beauty with Moral Considerations

3. Promoting Ethical Discourse

Open dialogues about the ethical implications of genetic beauty practices are essential for fostering a culture of responsibility and understanding. Engaging various stakeholders—including scientists, ethicists, policymakers, and the public—can help create a comprehensive view of the implications of these technologies.

- **Forums and Workshops**: Organizing public forums, conferences, and workshops can facilitate discussions that include diverse viewpoints. This ensures that a wide range of ethical concerns is addressed and that the conversation is inclusive.

- **Continuous Education**: As genetic technologies develop, so too should the understanding of their implications. Ongoing education about the ethics of genetic enhancement is crucial for professionals and the public alike.

4. The Role of Regulation

Regulatory bodies have a pivotal role in ensuring that genetic beauty practices adhere to ethical standards. Striking a balance between innovation and oversight is critical to preventing misuse and ensuring public trust in genetic technologies.

- **Creating Guidelines**: Regulatory agencies should work closely with genetic researchers and practitioners to develop guidelines that outline acceptable practices in genetic enhancement. These guidelines should be adaptable to evolving technologies.

- **Monitoring Compliance**: Continuous monitoring and evaluation of genetic enhancement practices are necessary to uphold ethical standards. Regulatory bodies must have the authority to impose sanctions for unethical practices.

Engaging in Thoughtful Dialogue About the Future of Genetic Modification

5. Cultivating a Culture of Ethical Reflection

Encouraging individuals to reflect on their desires for beauty through the lens of ethics can foster a more responsible approach to genetic enhancement. This reflection can help individuals understand the broader implications of their choices.

- **Personal Reflection**: Individuals should consider their motivations for pursuing genetic enhancements. Are these desires influenced by societal pressures, or do they stem from personal aspirations? Engaging in this self-reflection can lead to more informed and conscientious decisions.

- **Community Engagement**: Encouraging community discussions about beauty, ethics, and genetic enhancement can promote a culture of shared values and collective understanding. This approach can mitigate the pressures of societal beauty standards.

6. Future Ethical Considerations

As gene editing technologies become more accessible, new ethical dilemmas are likely to emerge. Anticipating these challenges and preparing to address them is essential for ensuring responsible use of genetic enhancement in beauty.

- **Evolving Standards**: Ethical standards must evolve alongside technological advancements. Continuous assessment and adaptation of ethical guidelines will be crucial in addressing new concerns as they arise.

- **Interdisciplinary Collaboration**: Collaboration between ethicists, scientists, healthcare professionals, and sociologists will be necessary to navigate future ethical dilemmas effectively. This interdisciplinary approach can provide a holistic perspective on the implications of genetic enhancements.

Conclusion

Finding balance between the pursuit of beauty through genetic enhancements and ethical considerations requires a multifaceted approach. By establishing clear ethical boundaries, promoting open dialogue, ensuring equitable access, and fostering continuous education, we can navigate the complexities of genetic editing in a responsible manner. The next chapter will explore in-depth case studies in genetic beauty, examining the real-world implications of these technologies and the lessons learned from their application.

Chapter 24: Case Studies in Genetic Beauty

In the rapidly evolving field of genetic editing and beauty enhancement, real-life applications provide crucial insights into the implications, successes, and challenges associated with these technologies. This chapter delves into several noteworthy case studies that illustrate the transformative potential of genetic modification in the pursuit of beauty, along with the ethical, social, and personal ramifications of these interventions.

1. Case Study: Skin Rejuvenation through Gene Therapy
Overview

One of the earliest and most publicized applications of genetic modification for beauty enhancement involved gene therapy for skin rejuvenation. Researchers at a renowned biotechnology company developed a treatment that uses modified skin cells to enhance collagen production, significantly improving skin elasticity and reducing signs of aging.

Implementation

Patients underwent a minimally invasive procedure where a sample of their skin was taken and genetically modified in a laboratory setting. The modified cells were then reintroduced into the patient's skin. Clinical trials showed promising results, with participants reporting visibly smoother and younger-looking skin after a series of treatments.

Outcomes

The treatment was hailed as a breakthrough in anti-aging technology, allowing individuals to reclaim a youthful appearance without the need for invasive surgeries. However, discussions around the ethical implications of manipulating genetic material for cosmetic purposes emerged, highlighting the need for informed consent and thorough understanding of long-term effects.

Lessons Learned

This case underscores the importance of transparency in the genetic modification process, particularly in communicating potential risks and benefits to patients. It also opens a dialogue about the societal pressures that drive individuals toward such enhancements and the need for broader acceptance of natural aging.

2. Case Study: Hair Restoration through Genetic Editing
Overview

A groundbreaking study focused on genetic editing to promote hair regrowth in individuals experiencing pattern baldness. Scientists utilized CRISPR technology to target specific genes associated with hair follicle development and growth.

Implementation

Participants in the study received a treatment involving gene editing to enhance the expression of genes that stimulate hair growth. Following treatment, subjects were monitored over a six-month period to assess hair regrowth and overall scalp health.

Outcomes

Initial results showed a significant increase in hair density and thickness, leading to positive self-esteem and confidence among participants. However, ethical discussions arose around the manipulation of genetic traits associated with one's appearance, with questions about how societal beauty standards could influence individuals' decisions to seek such modifications.

Lessons Learned

This case highlights the necessity of contextualizing genetic beauty enhancements within the framework of cultural standards and individual identity. It emphasizes the importance of fostering a culture that values diversity in beauty, thereby reducing the pressure to conform to narrow definitions of attractiveness.

3. Case Study: Body Contouring through Genetic Engineering

Overview

A pioneering approach to body contouring involved genetic engineering to alter fat distribution and metabolism. This experimental procedure aimed to provide an alternative to traditional liposuction and body sculpting techniques.

Implementation

Patients participated in a clinical trial where targeted genes related to fat metabolism were edited to promote a healthier fat distribution pattern. The goal was to enhance the body's natural ability to shed excess fat without surgical intervention.

Outcomes

The results were initially promising, with participants reporting favorable changes in body shape and composition. However, as the treatment progressed, some individuals experienced unforeseen side effects, prompting a reevaluation of the long-term safety of such genetic modifications.

Lessons Learned

This case stresses the critical importance of rigorous testing and post-treatment monitoring in genetic beauty applications. It also raises concerns about the potential for body image issues to be exacerbated by genetic modifications, underscoring the need for comprehensive psychological support for patients undergoing such transformations.

4. Case Study: Genetic Testing for Personalized Skincare
Overview

A leading skincare brand integrated genetic testing into its product offerings, allowing consumers to receive personalized skincare solutions based on their genetic profiles. This approach aimed to tailor products to individual skin types and genetic predispositions.

Implementation

Consumers underwent a genetic test that analyzed specific genes related to skin health, hydration, and aging. The results were then used to develop a personalized skincare regimen that included products optimized for each individual's unique genetic makeup.

Outcomes

The initiative received positive feedback from users, with many noting significant improvements in skin texture and overall appearance. However, the integration of genetic testing into consumer products raised ethical questions regarding data privacy and the commercialization of genetic information.

Lessons Learned

This case illustrates the potential of personalized beauty solutions but also emphasizes the need for strict data protection measures. Ensuring that consumers' genetic data is handled with care and transparency is crucial to fostering trust in such innovations.

Conclusion

These case studies illustrate the multifaceted implications of applying genetic editing technologies in the realm of beauty enhancement. While they showcase the remarkable possibilities that genetic modifications can offer, they also highlight the ethical dilemmas and societal challenges that accompany such advancements. As the industry progresses, it is essential to engage in ongoing discussions about the impact of these technologies on personal identity, societal standards, and the ethical landscape of beauty. The next chapter will offer concluding thoughts on the journey of genetic beauty and the future that lies ahead.

Chapter 25: Conclusion: The Path Forward

As we conclude this exploration of genetic beauty and the powerful implications of gene editing in enhancing attractiveness, it is vital to reflect on the journey we've undertaken throughout this book. We have delved into the scientific foundations of genetics, the revolutionary technologies that allow us to manipulate these foundations, and the ethical considerations that accompany such advancements. It is clear that the intersection of genetics and beauty holds tremendous potential, but it also requires thoughtful navigation.

Recap of Key Points

1. **Understanding Genetic Influence**: We began by establishing the fundamental role that genetics plays in determining physical appearance and its impact on societal perceptions of beauty. Genetics is not just about inherited traits; it shapes our self-image and interactions with the world.

2. **Advancements in Gene Editing**: Technologies like CRISPR have opened new avenues for cosmetic enhancements. The ability to edit genes offers the possibility of tailoring our appearance to align with personal ideals of beauty, challenging traditional approaches to aesthetics.

3. **Ethical Considerations**: The exploration of the ethical landscape revealed the complexities of modifying human traits for cosmetic purposes. We discussed the importance of informed consent, societal pressures, and the potential consequences of prioritizing certain standards of beauty over others.

4. **The Role of Environment**: The interplay between genetics and environmental factors highlighted that beauty is not solely determined by our DNA. Lifestyle choices, nutrition, and mental health significantly influence how our genetic potential is expressed.

5. **Community and Acceptance**: The importance of community support and dialogue around genetic beauty practices emerged as a recurring theme. Building a culture that embraces diversity in beauty can mitigate the risks of stigmatization and

The Future of Beauty in the Age of Genetics

enhance the acceptance of genetic modifications.

Looking ahead, the future of beauty is poised for transformation. As genetic technologies continue to evolve, we may witness increasingly personalized and effective beauty treatments that cater to individual needs and preferences. However, this potential must be balanced with ethical considerations and a commitment to responsible use.

The challenge lies in fostering a society that values all forms of beauty, encouraging people to embrace their unique attributes rather than conforming to narrowly defined ideals. The ongoing conversation about genetic modifications should not solely focus on enhancement but should also address the inherent value of diversity in beauty standards.

Encouragement for Responsible Advancements

As we move forward, it is essential to advocate for transparency in genetic beauty practices and to prioritize ethical considerations. Engaging in meaningful discussions about the implications of genetic editing will help shape a future where advancements in beauty technologies are guided by respect for individuality and diversity.

Individuals should be empowered to make informed choices about their beauty journeys, armed with knowledge about both the possibilities and limitations of genetic enhancements. Education will play a crucial role in dispelling myths, addressing fears, and fostering an environment of acceptance around genetic beauty practices.

Final Thoughts

In closing, genetic charisma is not merely about achieving an ideal appearance; it is about harnessing the potential of science to enhance well-being, boost self-confidence, and embrace one's unique beauty. As we stand on the brink of a new era in beauty, let us approach the future with curiosity, responsibility, and an unwavering commitment to celebrating the rich tapestry of human expression.

Together, we can navigate the complexities of genetic beauty, shaping a world where advancements in science harmonize with the core values of acceptance, diversity, and the profound beauty inherent in every individual.

www.ingramcontent.com/pod-product-compliance
Lightning Source LLC
Chambersburg PA
CBHW082110220526
45472CB00009B/2125